UFOs and Aliens
A Simple Guide to Main Alien Races on Earth and How to Contact Them

By

Alan Fredrich

UFOs and Aliens

A Simple Guide to Main Alien Races on Earth and How to Contact Them

Alan Fredrich

Copyright © Alan Fredrich

All rights reserved. No part of this guide may be reproduced in any form without permission in writing from the publisher except in the case of brief quotations embodied in critical articles or reviews.

Legal & Disclaimer

The information contained in this book and its contents is not designed to replace or take the place of any form of medical or professional advice; and is not meant to replace the need for independent medical, financial, legal or other professional advice or services, as may be required. The content and information in this book has been provided for educational and entertainment purposes only.

The content and information contained in this book has been compiled from sources deemed reliable, and it is accurate to the best of the Author's knowledge, information and belief. However, the Author cannot guarantee its accuracy and validity and cannot be held liable for any errors and/or omissions. Further, changes are periodically made to this book as and when needed. Where appropriate and/or necessary, you must consult a professional (including but not limited to your doctor, attorney, financial advisor or such other professional advisor) before using any of the suggested remedies, techniques, or information in this book.

Upon using the contents and information contained in this book, you agree to hold harmless the Author from and against any damages, costs, and expenses, including any legal fees potentially resulting from the application of any of the information provided by this book. This disclaimer applies to any loss, damages or injury caused by the use and application, whether directly or indirectly, of any advice or information presented, whether for breach of contract, tort, negligence, personal injury, criminal intent, or under any other cause of action.

You agree to accept all risks of using the information presented inside this book.

You agree that by continuing to read this book, where appropriate and/or necessary, you shall consult a professional (including but not limited to your doctor, attorney, or financial advisor or such other advisor as needed) before using any of the suggested remedies, techniques, or information in this book.

TABLE OF CONTENTS

Introduction..7

Chapter 1
What Public Thinks Of the Possibilities of Alien Existence?..8

Chapter 2
The Seven Different Known Alien Species and Their Origins..10

- The Reptilians..10
  - Reptilian Alien Description..........................11
  - The Reptilian: Their Planning and Agenda...12
  - Ancient Texts And Their Reference............13
- The Grey Alien..14
  - Detailed Description......................................16
  - Their Story Of Origin.....................................16
  - The So Called Agenda...................................17
- The Nordics..18
  - Description of Nordic Alien..........................18
  - The Nordic Planning and Agenda................19
  - Origin Theory...19
  - Conclusion..19
- The Zetas..20
- The Sirians..22
- The Orions..24
- Procyonians..26
  - Procyonians And Their Origin Theories........27

Chapter 3
Alien Contact **28**
   Change Your Diet 29
   Face Your Fears!...33
   Raise Your Vibrational Level...........................39
   Change Your Chemical Dependency! 41
Conclusion..**44**

Introduction

This is probably the biggest mystery of all the times on the existence of intellectual species other than human outside the earth. The fact remains undiscovered but the incidents and real experiences might tell you about something completely different picture. Although no evidences are conclusive enough to prove anything but then so many things are also not being proven and yet it is up to the individual to form their opinion.

Leading national governments are never in the favor of any existence but as we all know that they have secret projects and secret areas that deal in this kind of matter in the name of national security. Many people year on year reports about bizarre incidents and sightseeing that is not possible to carry out by any human efforts. People have confirmed such strange lights and certain way of appearance in the sky that is not possible by any military aircraft or any man-bound plane.

Many people very confidently believe to have seen the alien UFO but since it is just the observation by them and not any physical proof. "Me and my life has personally seen the socking UFO coming and that took our breath away for some seconds but we know that no one will believe us what we saw" told by a couple from Texas. Many skeptics does strong argument that evidence are not always look the way as all the natural incidents can be explained by natural activities. And no matter what people say the question of reliability always exist. It should not be confused with some abduction stories.

Chapter 1
What Public Thinks Of the Possibilities of Alien Existence?

The results speaks for themselves, a survey is being conducted to know the opinion of people about the reality and their perceptions about their likelihood. In the shocking result 60% of the people said that maybe there are existing in somewhere in the space but they don't believe that they are existing on our earth. In about 22% of the people said they even come to the earth and they firmly believe of their existence. In the survey 8% of the people have believed to have their presence in the movies only and 9% of the peoples said that they firmly do not believe and there is no such possibility of their existence. In that search SETI is actively searching through out though radio signal analysis by their patterns and intelligent communication. They are trying to find the noise pattern though searching through cosmic activities. In some incidents they confirmed to have received anomalous signals but not any kind of patterns that can explain the radio communication through signals.

By all that means the likelihood of their presence is showing many optimism and opportunity in the future to prove their involvement with the humans but to get any conclusive evidence or eye popping proof, the humans have to wait for many years to come.

Many technological efforts have been made in order to track any unusual activity in the space. The radio waves have been sent in the outer space till the eternity and many messages in the form of signals are also being thrown in the space, many trusted sources have also said to have received the responses that would indicative of such possibilities of life existence outside the space. In order to search for such possibility an equation is also being drafted by Frank Drake that calculates the chances of inter. Many expert scientists also argue that we live with the mental frame that it is uncommon in space for life existence.

On the other side of the question that Do Aliens Exist. There is question of the rare earth hypothesis. In one of the book, a physicist lists total of 33 characteristics a planet must tend to be supporting life. He estimates the probability of any such a combination that can be found in the universe is less than one in a million trillion.

In another book, an astronomer Guillermo and theologian Richards carry the notion further ahead by asserting that our place in the universe is not special but also designed for discovery of life.

Chapter 2
The Seven Different Known Alien Species and Their Origins

There are numbers of alien species that exist in the literature and many articles and information is being shared to about their existence, in the following article we take you to the world of alien species and how they are being originated. Some of them are described by people who claimed to have seemed them in real and many have formed the description on the basis of certain literature and references. Information about the most common species is being covered

The Reptilians

There are hundreds of theories about another race of aliens which is commonly known as the Reptilians. One being is that the Reptilian origins from the constellation of Draco. On the other hand it is also to be believed that ancient astronauts had seeded our earth long time back with Reptilian aliens. They did that with the main intention of creating a subset species to dwell around the earth.

Reptilian Alien Description

The earth Reptoid's facial features completely has different look and feel than typical and widely known greys. They have slightly cone shaped head structure and they have the bony ridges at the start of the forehead. From there all it runs all across way to the back of the skull. There are some slits in the nasal area. That assists them in breathing these slits are functions for them as ears also. Strange it is but this the way their structure is made. There are two different anatomy of eyes they both are large in size and some of them are black colored with pupils having vertical slits. It is just likes of lizards & snakes. On the other hand the eye's anatomy consists of white color with flaming slits for pupils.

Only about 20% of alien abductions account for being conducted by Reptilian. Based on these reports, the Reptilian that dwells on the earth are being described as humanoid. Their average weight is and about 6 to 9 feet tall. Their body is very muscular that resembles with that on Humans. They have long arms and legs, and each of the hand has 3 and sometimes 4 long fingers and one thumb, just like human. Their skin appears dark green with brown scales. Many also reports of them with having tails also.

The alien Reptilian is said have real origin from Draco. Draco differs from Reptoids in many fronts, which is by the way their earth dwelling relative. Apart from their eyes, their total height is also ranging from between 6 and 9.5 feet. Reptilian also has wings that are made of bone which is covered with scales of 6 to 8 feet. Their body structure resembles of earth Reptilian except for the part that they have been seen with horns on their head, sometimes along the length of their spine also. These body characteristic remember us of dragon and many winged god that are being found in the historical literature of Greek and Indian origins.

The Reptilian: Their Planning and Agenda

Reptilian abductees described their nature of intimidation in order to extract information from human captives. They made the human captives do things by the means of creating fear and force. Conspiracy theories state that these Reptilian is a malevolent group which seeks dominion over all around the universe including out mother earth. Their dominant nature is visible from their act of cloning Grey aliens so that they can use them as their slave. Similarly they have the same plan with humans also. There are many strong believers in their intention of slaving the humans and they also believes that they live in the 3rd, 4th and 5th universal dimensions. They feed themselves on negative energy and in order to fulfill their dreams they creates conflicts in our world.

One British origin writer firmly believes that they have successfully created a mix breed of human and reptilian that is can be referred to as of "elite" class. These super intelligent species are now in power and infiltrated all government units, heavy corporate, banking sectors and even filthy reach families. They are capable to adjust their shape to hide their true identification. They have one and only bad intention of taking over of the planet by causing conflict and war around the world. They want to do that to create negative energy from misery by corrupting the humanity to fulfill their bad intention. The earth dwelling Reptoid species lives deep underground and in caves where no one can track them and continue planning their evil agenda.

Ancient Texts And Their Reference

- **Chinese-**Dragons had the ability to change their shapes and turn into humans. They superior species are the dragon ruler of the all four seas – Be it north, west, east or south. They style of attacking the people are without provocation and the temple of honor is also built for these kings in order to appease them.

- **Greek-** Cecrops is the believed to be the first king of Athens which is a half snake in the nature. In the Pergamum temple, the sculptures are prevalent that contains men with the serpent legs. Boreas was the Greek god of the north and which has the serpent legs and wings.

- **Indiacontext-** Indian scriptures describes the Naga legend, which is also reptilian being that frequently communicates with the people and lives underground. In another term they are referred as Sarpa also.

The Grey Alien

The most known of alien kind, they are typically described as a thin, very thin actually as compared to human, being with an oversized head in comparison to their total size of body. They stand three to three to six feet taller. They weigh from 55 to 95 pounds in between.

The Greys have been creating problems for your government ever since their first downed craft was discovered in the New Mexico desert in 1947. The creatures you found (for that is what you considered them) both horrified and mystified you. The authorities were immediately notified and the few civilians present were ordered to be silent. The Air Force whisked away the remains of the fallen craft and its inhabitants, one of which was still alive. A special task force was organized to examine the findings and give recommendations. The one surviving extraterrestrial was taken to a laboratory and kept under strict supervision day and night, where after some years it succumbed to an unknown illness. This was your first encounter with the Greys.

Other governments have had some encounters with the Greys, but none so far-reaching or complex as yours. The United States is the only government that has had ongoing communication with the Greys, including a treaty, which has turned out to be a sham. Other governments have merely encountered the Greys and other extraterrestrials briefly, and kept reports on these meetings and other sightings.

The Greys are more interested in the United States than in other countries because it sees the United States as the most powerful nation. They are right about that, but they have underestimated your intelligence and your love for freedom. If it were not for their advanced technology, you would be a formidable foe. Their technology is not nearly as powerful as they would like you to think, however.

Your government doesn't know what to do about the Greys. It has had relations with them for over forty years, during which time several administrations have come and gone. Each administration during those years has known something about the Greys, the treaty with them, and the history of your relationship with them. However, over the years, continuity has been lost, facts have been confused, and now no one knows the entire story. This along with the Greys' lies has made this situation less and less manageable over the years. The ones in charge of this no longer know what the truth is, since none of them were around for the initial recoveries and treaty. What has happened is a little like "Pass the Secret," where no one in the end knows what the secret was.

This is fine with the Greys. The longer they can keep their activities secret and keep you in a state of confusion and division, the better. They are trying to back your government into a corner so that it will do something drastic, like detonate a nuclear device. By locating in the United States, they hope that you will destroy your country while trying to protect yourselves from them. They don't want the government to tell the public about them because they believe they can incite those heading this secret operation to attack them.

Detailed Description

In the detailed description, they have most commonly appeared in and covered by fiction novels and various literatures. Most commonly they are part on majority of the movies being filmed and they are internal part of the script when any one thinks of movie on alien.

They are generally being described as large and bulbous forehead that bulges, having unblinking eyes. Their Skin is so pale and grayish, plus has scary wrinkles on it. For the one who have encountered these types of creatures characterize their skin as of comprising to have the texture of an ocean mammal, just like a dolphin or may be whale. They don't have ears or indentations on the any of the areas of head where something to hear is can be identified. Many people have described them as having 4 to 5 fingers, with webbed hands and also webbed feet.

Their only and major method of communication is via telepathy. Most of the previous fantasies are indicative of talking within their mind in order to communicate to each other.

Their Story Of Origin

The origin of this type of alien is coming from a star map that is being drawn from abduction stories brought back during the hypnosis of the famous Betty Hill. UFO researcher based on analysis information came up with a hypothesis that their home base is a pair of stars which is known as "Zeta Reticuli". These stars are located at the farthest of 220 trillion miles from earth.

Sometimes Grey alien is also being known as the Zeta Reticulans and Reticuli

For the most part during their life, the Grey alien is scientifically observing other life forms in the universe including earth. They perform different tests and medical analysis upon human subjects and then free them after they do their research.

The So Called Agenda

The alien is dying because of originally they are cloned from reptilians. Originally Reptilians apparently created the Grey alien and then they forced them into servitude. That continued till the Grey aliens formed a rebellion. After the rebellion they escaped into the universe to wander around and by wandering they landed in earth and made earth as their home. Some theorists also believe that the Greys are originally from Roswell and New Mexico. Plus they have made an agreement with the U.S. government also. They offered advanced and latest technology and in exchange they have asked to allow them to experiment with human subjects. These experiments are specifically based on genetic engineering so that they can find the solution to safeguard their race. This is the main reason that the majority of alien encounter fantasies are about the greys.

The Nordics

The Nordics are described humanoid extraterrestrials that are also being called as "Blondes". They have a human like resemblance. Their average height ranging from five and a half feet and max it reaches to 7 feet.7

Description of Nordic Alien

They mostly have blonde hair and light eyes. Also their most typical of eye color is of blue. However many reports had described their other colors of eyes also. Most common are pink, violet, purple and green. He Nordics are known to be in perfect physical shape that can attract humans easily. And can easily resemble to human from the distance. Mostly they communicate with the techniques of telekinesis and telepathy.

The Nordic Planning and Agenda

Nordics are silent observers of Earth, they analyze every human activity and important changes in the atmosphere. Their sole intention is to enlighten by providing revelations and warn people of the earth about their different behaviors which can lead to catastrophe that includes issues surrounding many actions done by the Greys and Reptilians. Some proponents states that ideas of the New Age movement were provided to us by them only.

Origin Theory

Originally they originate from the Pleiades star cluster which is at the distance of 400 light years. Some of the very first stories about them have been written by Billy Meier. He claims he had several visits to Nordics in around 1940s. All of his information derives from discussions carried out in the late 1970s and also early 1980s. These discussions are made with a female Pleiadian. As he describes to him, their home planet is named "Erra". It has different dimension than ours. Nordics are an ancient humanoid species that discovered earth in way back in around 225,000 B.C.

Conclusion

It is clear that the information concerning all these alien beings whether Greys, Reptilians or Nordics looks like reading a good Sci-Fi novel!

Those who have encountered them are very reliable in their testimonies about their experiences. Many also come from normal backgrounds with good families. These families are having high education, work and normal physical as well as mental medical histories. There are thousands who have had some kind of alien encounter so these experiences can't discount their stories as untrue.

The idea of unknown aliens visiting on earth and abducting looks a bit far-fetched and exciting for others. By bringing clarity and understanding to this situation helps all in our quest for knowing what is happening in our world.

The Zetas

The history of this race is not only interesting but important to you. It is important because you are traveling down the road they traveled many millennia ago. They regret they traveled this road and have much to teach you about where they went wrong and what they would have done differently. You are giving them the chance to redo their past by providing them with the genetic keys to go backwards in time. You are also giving them the opportunity to balance their mistakes by teaching you. This is your relationship with the Zeta Reticuli.

The Zetas only recently became aware of you. They had never interacted with the human race before their arrival in mid-century. They heard the call for help sent out by your extraterrestrial guardians at the advent of the atomic age. When civilizations reach this point in their development (and many inevitably do), warnings are sent out to those who might be affected. The Zetas received this call because it was thought they might be able to convince you to turn away from nuclear power. The Elders of your planet, the Sirians in particular, saw the possibility of mutual aid and told the Zetas about you.

The Zetas were elated to discover there was a race so like them genetically and one to which they could pay their karmic dues. They eagerly accepted the challenge of helping you and, in the process, of helping themselves. The only problem was that approaching you directly would probably create too much disturbance on Earth. This had to be handled carefully. So the extraterrestrials guarding your planet devised a plan.

Their plan was to introduce themselves gradually to you, to secure soul-agreements with certain people to help, and to save genetic material in case you destroyed yourselves and your environment with nuclear weapons before they could stop you. They also planned to use the Pleiadians as go-betweens and peacemakers.

The Sirians

Already quite a bit has been said about the Sirians, since they played such a pivotal role in the human race's early development. They are still involved in your world today. The Sirians are guiding you through your intuition and dreams, through channeling, and through visions. Visions are the most direct way they have been involved with you lately, since they do not materialize or walk among you in physical bodies any longer. There are some reincarnated Sirians among you, but none in materialized bodies.

The Sirians do not have physical bodies, although they can materialize them if they want. As fifth-density beings and beyond, they function beyond matter. This has always been so and one reason they have seemed like gods to you. You saw them materialize and dematerialize before you, something you assumed only a god could do. In addition, many of the forms they took were brilliant, beautiful, and light-filled, making them all the more god-like. Occasionally, they still appear as beautiful light-beings or angels.

Your ancient representations of these light-beings had wings to symbolize their ability to appear, disappear, and move through the air. However, they did not appear to you with wings at first. Only later did they sometimes have wings, in keeping with your image of them. Thus, angels with wings became part of your mythology. But they did not always appear as angels. The form they took depended on who they were appearing to and why.

Although other extraterrestrials also appeared in visions as various characters and as angels, this role fell largely to the Sirians because they were the ones most involved in the human race's development and evolution. The Sirians acted as teachers and guides more than any other group, although others often helped them. Sirians have been particularly involved in your religious rites and religious societies, including all the mystery schools.

In earlier days, religion was at the center of society. From religion came the ethics that helped societies cohere and function. Laws evolved from these ethics, as did customs, rituals, and traditions. The religious leaders that headed society received guidance directly from the Sirians, so shamans and priests served as bridges between the unseen world of the gods (the Sirians) and the physical world of their own kinsmen. Sometimes these religious leaders were even reincarnated Sirians. Societies have always had members who could communicate with—or channel—these realms, and the Sirians made good use of them. This continues today, of course.

The Sirians described so far were positive, or at least they intended to serve you even if they sometimes fell short. However, throughout history, there has been another Sirian influence on your planet that has not been so positive. Service-to-self Sirians, negative Sirians, have played the role of Satan and Evil on your planet, with the help of some others. Like the positive Sirians, the negative Sirians have materialized and dematerialized as well as reincarnated among you. And through channeling and other psychic means, they have delivered their own version of truth.

The Orions

The negative Orions like to think they are important to the negatives' mission to capture the Earth, but their importance and impact are minor. Hence, the brevity of this chapter. The negative Orions are the most visible of the service-to-self extraterrestrials, usually taking the form of Men in Black. Others have arrived as Star People and Walk-ins. Thus, the negative Orions are involved with Earth in all the ways already described: through materialization as Men in Black, as Star People and Walk-ins, and through channeling and other psychic means.

Since negative Orions are third, fourth, and fifth density, not every Orion who reincarnates here would be considered a Star Person. Third density

Orions would not. Most of the negative Orions on your planet are reincarnations and fourth-density Star People and Walk-ins, making the Orions as a whole the least evolved of the extraterrestrial groups visiting you. Men in Black, who are fifth density, are rare, although quite visible.

Men in Black are fifth-density service-to-self Orions who temporarily take on a body through materialization to achieve their negative goals. Thus, they can affect matter when they need to and dematerialize when it is no longer expedient. The bodies they take on are usually of male gender dressed anachronistically in a 1940's style dark suit and hat. They look pale, sickly, unattractive, and emotionless. They impress people as extremely odd, not only because of their clothing and lack of emotion, but because of their whole demeanor. They have not studied your culture sufficiently to imitate you well, so they seem strange to you.

The Men in Black are aware that they seem strange to you. That's fine with them, as they are not here to win your favor or even to fool you, but to go about their business. Your reactions rarely interfere with this, so they have not bothered to refine their act. In fact, they like startling and scaring people. After all, this is something self-servers enjoy, so they are not the least bit concerned about what you think of them. The more mystery and fear they can create, the better.

However, mostly, they are just trying to get their work done.

The Men in Black are following a plan. They want you to believe that anyone who claims to have seen a UFO is being spied on by your government. Consequently, they pose as government agents and threaten those who make such claims. They are not trying to cover up evidence of UFOs; they just want you to think your government will do anything to keep this information secret. It's not hard to leave people with this impression, since your government has, indeed, been running a UFO cover-up. That is what makes the Men in Black's job so easy. The Men in Black are drawing attention to the government's cover-up by acting so strangely and by threatening people.

The reason they want you to think your government is intimidating people is that they want you to believe your government is a threat to its citizens. They don't want you to believe in your government, or in democracy for that matter. Democracy is counter to any political system they would ever create, since they believe in subjugating people—which is what they hope to do to you.

These Orions come from a system (in the past) with only two social roles: rulers and slaves. They believe they have the right and the duty to subjugate people and that those who succeed deserve to rule. They also believe that those who don't succeed deserve to be slaves. Slaves in the

Orion system also believe this, for this philosophy has pervaded their thinking for generations. So the oppressed have internalized their oppression so fully that the rulers have to do very little to keep them serving them. This leaves two classes of people: perpetrators and victims.

Procyonians

It is considered to be the scariest extraterrestrial groups are in our solar system which revolves frequently around Procyon.

Procyonians And Their Origin Theories

Procyon is a star that has the yellowish or white and rises before Sirius. The total distance from the earth is 11.4 light years. They has the Swedish nick names and they appearance has the blond hair. The US organizations are not at all convinced to talk and negotiate with them as they don't believe that they would let us give their techniques and weapons. In the many revolutionary developments the Procyonians is developed as a cross bread. Procyonians are not harmful to humans and they maintain very positive attitude towards human mankind. They many times protect the earth from the devil activities of greys and also Reptilians. They live with the philosophy to service human kind without expecting anything in return.

In the special abilities they can travel though time and also in dimensions. It is their most common used vehicles used to travel but not always dependent of them. The translation of Procyonians is something like this "The home for the ones has the ability to travel in time". They don't interfere in our activities nor do they create any imbalance on our atmosphere. They understand the free choice of humans.

Chapter 3

Alien Contact

Without following many of the techniques in this book, remembering a contact with an alien is fairly unlikely, unless that was already a planned part of your life path - an appointment organized before you were born.

There are a few things you need to know before exploring alien contact.

Why do you want private contact with aliens?

Firstly, and this is very important, please consider WHY you want alien contact. If it is because you think it is an exciting thing to do, that you have something you'd like to offer aliens in exchange for them meeting you or you want to make a great contribution to galactic society, read on.

What's the great law of the universe?

Secondly, in our universe, like attracts like. If you're unhappy, you'll attract more unhappy situations to reinforce your unhappiness. If you're happy you'll attract more happy situations to reinforce your happiness.

How does fear affect your connection with aliens?

If you have any fears about it, if you believe that aliens are only evil and always want to invade and take over, then, in rare cases, you may attract a negative alien. That may sound a bit scary but even aliens know that there is only positive and negative energy and both are valid. But, interestingly, most negative energy aliens are by themselves, but most positive energy aliens have worlds behind them, so it is much easier to meet a positive alien than it is to meet a negative one. The techniques in this book will allow you to be more likely to attract a positive alien. And, now that the human race has advanced sufficiently that Official First Contact is imminent, negative aliens are having a hard time even trying to locate our planet, let alone a negative person on it. Our advanced level is now making it harder for negative aliens to even make a connection.

Change Your Diet

We can make time for meditation, we can change our viewpoint on things to be happier, we can face our fears step by step, but changing the food we love or the eating habits we have or even the location we live in to be able to access different types of food can be difficult.

But what you ingest affects your health, your well-being, and your vibrational level, and increasing your vibrational level is extremely important for being able to meet aliens on their own level.

Actually, aliens usually only absorb into themselves things that nourish their energy level and their frequency and keep them at their high rates of ecstasy. Many do not eat but simply absorb energy directly from the environment. They understand that it is almost impossible for most human beings to be able to do that as our bodies may need decades to adjust.

But you can still make some changes to your diet to be able to connect with aliens.

Firstly, everything you eat has a frequency and what you eat also attracts more of what you eat. Beer, for example, is a low frequency drink and reduces your vibrational level. If you drink a lot of beer it can cause damage not only to your organs but also degrade your DNA. Not only that, but if your body is full of the frequency of beer then your body will signal that it requires other things of similar frequency to match. And so you will feel hungry for junk food, fatty food, food with lots of sugar in it and other bad food that has a similar vibrational level to beer, reducing your frequency further.

Meat is another frequency lowering ingredient. Excessive meat can not only increase cancer cell production as well as depression and short temper, it can also slow your entire system down, making you sluggish and unfocused. (Or focused on things that don't align with your core vibration)

Now I'm not against eating meat. Animals are second density creatures and have chosen to be here to help humans advance, as well as be a reflection to humans. Eventually humans will not eat meat in the future, but our bodies, our technology, our economy and our belief systems need to evolve first before we can all move to vegetables only and then energy only functionality. We've got centuries to go yet. But if you want to raise your vibrational level quickly, reduce your meat intake, slowly replacing it with fish, then replace the fish with vegetable substitutes until perhaps only one meal a week contains animal product.

This will help purify your body and make it easier for aliens to meet you.

So, in summary. You are what you eat, literally! Like attracts like. The more chocolate/meat/fat you eat, the more you will want to eat it. So take control of your diet and take control of your cravings and raise your vibrational level.

Exercise

List What You Ate

Make a list of everything you ate in the past week. This means even when you sprinkle something with salt or add a dab of mayonnaise. List everything.

If you have a loyalty card with a supermarket chain that has recorded your purchases over past years, you can refer to that too and make a list of things you've eaten generally.

Now look at the list and divide it into two columns. Those types of food that have high vibrational levels and those types of food that have low vibrational levels.

High vibrational level items include anything that has not been processed or canned, such as fresh fruit and vegetables that haven't been frozen or irradiated. Low vibrational items include anything long dead, preserved, mixed with other things, including meat and sugar, or packed in plastic. Medium to low also includes products that are produced by animals, such as milk, butter, cream, cheese etc. due to all the processes they go through before they get to you. This also includes anything alcoholic. Even though wine has a higher vibrational level than beer, alcohol in general is a low vibrational level liquid. Best avoided.

Now that you've created a list of things you eat and drink, look at the list of things you eat and put a line through things that you can replace with things with a higher vibrational level. Over the course of the next 30 days, slowly replace those items with higher and higher vibrational level items in your diet. If you already eat healthily generally, you can probably do that sooner. If you have never eaten healthily, it might take you longer than 30 days to do this.

Warning: Do not just throw everything out and start with just fruit and vegetables. Your body has adjusted to the food you eat and it'll take it awhile to grow into your new diet. Dramatic change could make you feel very sick and interfere with your work or other responsibilities. Take it slowly!

Examples of what you could do to raise the vibrational level or frequency of your diet.

Low Frequency

Apple Lollipop

Beer

Canned Pea Soup

Fries

Medium Frequency

Candy Apple

Quality Wine

Frozen Peas

High Frequency

Apple straight off a tree

Fresh grape juice

Peas straight off a plant

Potatoes just dug up

So, see if you can slowly change your diet to include higher frequency food types. Your body will then adjust to the higher frequency and you won't actually crave the lower frequency foods anymore. Interestingly, even if you try to eat the lower frequency foods at a later date, they actually won't be as enjoyable as you remember them because the receptors in your brain for that taste and enjoyment would have been replaced with enjoying higher frequency foods.

Face Your Fears!

The most important thing you can begin to do to be ready for your private alien contact situation is to face as many of your fears as possible.

This is not as hard or as daunting as you may think. We all go through life creating new fears and accumulating long lists of major and minor fears, but when we face all the major fears, things get easier and all the minor ones seem to drop away or fall down like a line of dominoes.

For example: You may have had a belief about something that a friend told you a while ago. When you go back along the memory, you realize your friend told you this... when they were 10! So, you should reconsider the advice now based on your increased knowledge and wisdom (and access to Google!) and see if it is still relevant.

Most of the time, things have changed, and it is a belief that you can forget.

Today you need to take an hour out of your time and write down as many fears as you can think of. When you start challenging yourself and writing down all your fears (start with fifty of them) you'll actually realize, just looking at what you've written, that many of them have been fears you've been holding onto since you were a child, and are no longer relevant. Some fears you think you have you may have already overcome without realizing it. Not until you take some time out and write them down will you be able to realize that you actually don't have that many fears these days.

So, what kind of fears do most people have?

Many are scared of insects or small animals. Many more are scared of open spaces or enclosed spaces, heights, confrontations, germs and more. Many of these are common across cultures and, as a result, there are actually training programs you can join that help you to get over these kinds of fears. i.e. you don't need to do it alone.

So, your first act on the first day of your new plan is to find out what fears you have and begin to deal with them. You don't need to deal with all your fears on the same day. It took me years to challenge myself to get over spiders and even now, walking through a web has me grimacing before I quickly calm myself and move on. While holidaying in Edinburgh I attracted the situation into my life where I happened to be at a butterfly sanctuary on the day that someone had come in with tarantulas for children to play with. I could have walked away from that but decided it was time to face my fear of spiders and hold a tarantula. (Admittedly a tarantula isn't really a spider but it's big and furry and has eight legs so the archetype is what I had to get over my fear of) so I held one and its legs felt like little cotton buds on my palm. I went from fearing it to finding it cute in moments!

There are other fears I quickly got over. Fear of heights is a very common one and I found my fear wasn't of heights, it was of doing something stupid while at a height. So I had no problems if I knew there was glass or a fence but if I was on slippery rocks at the edge of a cliff then my fear wasn't a fear at all, it was simply a warning that I needed to take more care. So I integrated that and moved on.

You will find the best way to get over your personal fears could be reading other people's examples of how they got over theirs, or joining a support group, or even having your friends help you out. Find a way to get over the major ones and the minor ones will quickly disappear.

Many of us have minor fears such as will people think I look big in this dress or if I grow a beard how will I be received in my workplace? There is a word that many guides, gurus and self-help enthusiasts use in this situation:

"SO?!"

Think about it. Do you really think your life should be controlled by what other people think? That will quickly pull your confidence down. If you are constantly worried about what your friends or your family or your work colleagues think of your appearance you will be fighting a losing battle to the end of your life, as you will always find someone who has an opinion on how you dress in your entire life.

As I'm a guy I'll give this analogy. Apologies that I'm unable to give a similar analogy for a woman, but I hope you'll get the essence of the message. If you're a guy you'll be worried about your spots in your teens, your muscles when you're in your 20s, your fat in your 30s, your hair in your 40s and your wrinkles in your 50s. By your 60s, if you've lived that long worrying about your appearance, you'll probably be worried about how your health is perceived by those around you. And as you age and meet new people they'll all have their opinion of who you are and what you look like.

It doesn't matter what they think! What matters is who you are and how you are pursuing your excitement in life. Don't let others dictate how you decide to look.

So, minor worries can be dismissed, but sometimes you may need to do a bit of work to dismiss them.

The best way to get rid of a lot of minor fears is to increase the fear level to the maximum it can be in your head, and then look at what the most extreme case is and what could happen next.

Minor Fear Example 1: *Speaking to the boss about a problem.*

You fear that speaking with the boss will cause a worse problem. Worst case scenario is that you lose your job. How terrible. How scary. Or, how exhilarating, what freedom! You could look for a new job or you could start your own business or you could take a break from work altogether.

After considering the fear you realize that even the worst case scenario is a positive one, and you go and speak to the boss about a problem without worrying about what happens next, as you don't care, it doesn't matter and it's nothing to be worried about. The most likely scenario is that the boss appreciates your feedback and fixes the problem!

Minor Fear Example 2: *Asking Someone Out*

There is a nice girl or guy you want to speak to as you are interested in them in a romantic way. But you're terrified they won't be interested in you. What's the worst case scenario?

The worst case scenario is that this person is so not into you that they turn into a screaming, crazy person full of insults and other terrible behavior patterns. The worst case scenario is that your image of them is completely destroyed and that you have no interest in them anymore!

As there are billions of people in the world and millions of them are available online, the likelihood you could find someone much better than them is very high, so that should reduce any fear you have of speaking with them in the first place. If there's nothing in common, it was a great fantasy for a while, and there's someone better just around the corner.

Having said that, people love confident people who go after what they want. If you go and approach this person confidently with no expectations, you'll probably find that you can have a great chat with them and may end up having coffee or a drink with them sometime down the track. And if you don't find that it happens that way, you still now have the confidence to approach many others that might be more suitable for you.

Minor Fear Example 3: *Meeting an Alien*

How will you react to aliens when they appear? Yes, it's a minor fear. There has yet to be a news report of people being killed by an alien. It's impossible. Even those who disappear for appointments usually return, unless they had decided to leave by choice, so going off with aliens is a minor fear.

However, you need to deal with this if you wish to be invited aboard a ship, so you need to consider the worst case scenario. My personal feeling is that getting to travel aboard a spaceship is worth a medical examination to start with and if any aliens want to have their way with me I'd be quite okay with that too. If they need to take samples, go for it. On Earth, many give gallons of blood at blood donor banks, and containers of urine to doctors, and all we get in exchange is maybe a cup of tea and a biscuit, or a bill. If aliens want some blood and urine in exchange for a tour of their ship, well, I'm in!

The situation can be very scary though, so you need to be able to put yourself in a situation where you accept your adrenalin rush and you accept that you will feel fear and you accept that it will feel like you're not in control as your fear is taking you over. If you can accept and integrate all your fears, knowing that many are simply your body alerting you to things you need to be aware of, and sort of fall into the fear and take it as just another feeling, then when aliens come to visit you won't be paralyzed and you might even be able to sit up and say hi.

I've heard from a couple of people who have had private contact that there is a grey alien that likes to drop by wearing a red cape. His idea of humor. Don't miss the jokes aliens try to make to cheer you up while you're paralyzed with fear. They might not be able to smile like we do but it doesn't mean they don't have a sense of humor! Look for the funny side.

So, to get over your minor fear of aliens, imagine you're in your bed, you can't move, you can't scream and all you can do is breath heavily with great fear. What's happening? Adrenalin rush and paralysis. Is it bad? No. Will it go away? Yes. What if there are aliens there? Do you want to go with them? Yes or no. If no, they'll go away. If yes, they'll wait for you to calm down a bit then take you with them. If you can get over your fear and say 'yes', woohoo! Trip on a spaceship for you! If you can't, accept it and do something in your life to prepare yourself for the next time.

Sometimes aliens will just appear briefly as a test, even as a shadow on the wall or even more subtly, like outside in the yard where you can't see them but you can sense them, and quickly disappear if they feel their presence is too frightening for you. So, don't worry. You're not going to have a heart attack if an alien appears and you're not ready. They won't let that happen!

So, how do you begin reducing your fears step by step so that you're ready in 30 days to remember your encounter?

Exercise

Morning Meditation

Get into a comfortable position. Close your eyes and think of the first minor fear you have.

Now bring that fear into focus and visualize everything you can think of associated with that fear. How does the fear manifest? What causes the fear to appear? Where does it come from? Why do you have that fear in the first place? How did it develop? Is it an old fear from childhood that no longer has any relevance or is it a new fear based on a lack of knowledge about something?

Sometimes simply not knowing the details of something can cause us to be fearful and so simply finding out the information can make the fear disappear.

Now that you have the fear in your mind and you understand where it has come from, build the fear in your mind until it goes to the worst possible place ever. Take it to the absolute extreme. What is the worst that could happen? And then, when you've worked out what the worst is that could happen, work out what will happen next after the worst that could happen, happens.

Remember that you are an eternal being and cannot be destroyed. You can transform into different forms but your spirit can never be annihilated.

One you've identified the fear and taken it as far as it can go, it's time to integrate it.

Raise Your Vibrational Level

What do people mean by raise your vibrational level? Well, we are all different kinds of frequencies. A metal detector works by detecting the particular frequency a particular metal gives off. Just like metal we all give off a frequency.

However, in third density/early fourth density reality on Earth, our frequency is very low and it can sometimes mean aliens don't even notice us without complicated equipment, and we certainly don't notice them.

The skies are a hub of activity. Thousands of UFOs are crisscrossing above, through and under our cities every day. But many of us are not at the right frequency to even detect them. Those that are will stand next to people who are not, pointing and shouting at an amazing light show while their friends are peering skyward unable to see anything.

An analogy would be a dog whistle. The frequency is too high for many older adults to hear and so it might as well not even be there. To many it doesn't exist.

But dogs can hear it. And so can children. If the dog whistle sound was the frequency for seeing aliens, and if you were to change yourself to be on a level where you could also hear it, then you would also be able to see aliens.

If that's a bit confusing, here is another analogy.

An alien is like the blades of a fan at full speed. The vibration of the fan blades, or rather, the frequency, is moving so fast that it looks like the fan has disappeared from this reality altogether.

If you could raise your frequency to match the fan blades, you'd be able to see them again, even though they're moving fast.

So, how do you increase your frequency?

There are many things that can change your frequency. The main thing though is your emotional state. When you are in a happy, joyful and confident state your frequency is actually higher. When you are in a depressed state your frequency, or vibrational level, is lower. You need to work on keeping your frequency high. The goal is a high frequency 24/7 and then to keep increasing it. There is a reason that many caring people in history are pictured with halos behind their heads. They'd increased their frequency to such a level that they had begun to glow. Actually, meta-physicists believe that we've already increased our vibrational level over the past two thousand years to such an extent that if we went back in time, everyone two thousand years ago would perceive us as glowing! So, that's your plan within this time period!

Here's an exercise to help you get started:

Exercise

Raise your vibrational frequency

To raise your vibrational level, your frequency, you simply need to increase your happiness factor. That's it. But if you're not a happy person to begin with, this may be a bit difficult for you.

By changing yourself to be happy at just about anything, you effectively raise your vibrational level so that not only does the world around you change to be more exciting and joyous, but you'll also start attracting more enjoyable things into your life.

To increase your vibrational level quickly, you need to see the world through the eyes of an incredibly grateful person. Be grateful that you're walking around a billion dollar city. Be grateful that you have food, shelter, friends, a job, air to breathe, sunlight to help things grow and more. Work on finding out all the things you're grateful for and then go over the list being happy about everything.

You can also do this at the end of your daily meditation. Think of the most amazing things you're grateful for and feel the emotion that goes with that. Feel how it seems to pull you up higher. Feel the happiness and gratitude fill every part of your body.

Change Your Chemical Dependency!

It can be completely shocking when we find out how many low vibrational level chemicals we breathe or ingest moment to moment. We never give it a second thought but every second of every minute of every hour of every day we are immersed in a toxic chemical soup.

Think about all the chemicals you encounter from waking up in the morning to going to sleep at night.

It's a wonder we still live. But one of the reasons why we're able to exist in this poisonous world is that generations upon generations have slowly adapted to the increase. Since the industrial age began, pollutants have been entering our atmosphere and we have been evolving along with them. But it doesn't have to be that way, and the best thing for us, to be able to continue to increase our vibrational level, is to remove as many of the toxins around us as possible, and thereby change our bodies to be able to resist them.

For many people who suffer from cancer or some other debilitating disease, removing the toxins from their environment would enable their immune system to get stronger and not spend its time fighting them, and then be able to combat the cancer cells in the body.

Remove the toxins, increase the immune system, increase the vibrational level and increase your likelihood of meeting some aliens.

Paradoxically, once our bodies are pure and at an extremely high vibrational level (think as high as monks in monasteries in the mountains for example) we reach a stage where toxicity is less likely to affect us. Everyone has a friend or acquaintance who knows someone who's lived to almost 100 smoking and drinking every day, even with doctors telling them for years to quit smoking or they'll die. Well, these people are extremely positive with great diets and their bad habits were less likely to affect them.

But if you're not at their positivity level, and you'd like to add meeting aliens to your list, you need to go for the full de-tox and regeneration option!

So, how do we do that?

Exercise

List the Chemicals You Use

Just like with the diet exercise, in this one you should write down all the items that you use every day that are not natural and see if there is a natural alternative.

There are some things we just have to live with, like fire retardant chemicals on every piece of furniture on the planet, but there are other things we can do to replace some toxic substances.

Instead of chemical soap, just use plain soap.

Instead of chemical toothpaste, just use baking soda.

Instead of aluminum based deodorants, use natural deodorants

Many people around the world have been finding non-toxic replacements for a wide variety of chemicals. You'll be able to find lists online.

Find what you can replace, replace them, and you will increase your vibrational level. You may also see your health increasing too.

Now, if you live in an area that is heavily polluted and no amount of buying healthy and natural replacements is going to change that, then the next thing to do is to change your location. Think about that on your review day.

Conclusion

Thank you again for downloading this book!

I hope this book was able to help you to effectively understand more about extraterrestrial species and aliens contacts

Have you had an experience with aliens yet that you've brushed off as just a dream? Or tried to convince yourself it didn't really happen? It really happened. Get that dream clearer and see if you can bring more and more of it into this reality. By following the steps above and increasing your vibrational level so that you are connecting more closely with aliens, you will be able to remember your experience with aliens more clearly, and be ready for their next meeting with you.

Initially aliens are likely to meet you in a co-created reality. It is like a bridge between our reality and their reality. As it is a much higher vibrational level than you're used to, your brain is unlikely to be able to bring much of the experience back with you and anything you do remember will feel like a dream.

The steps above will make these experiences more real, more palpable, even to the point of being able to still feel their touch, smell their smell, remember the conversation and know what you were helping them with at the time.

If you still haven't had an experience with an alien that you can remember, go over the steps above and see if you're missing something. Do everything again as thoroughly as possible. If you really want to meet aliens to help them with their work, don't give up. It'll happen for you.

For those of you who have had an experience with an alien, you may wish to read through and refresh yourself on some of the steps to help you make a stronger and more regular connection.

If you have gone over the steps several times in detail and are absolutely positive you've done everything possible and you're still not getting any results, either your life path is not ready for this experience, or you might be better suited to meeting them another way.

Thank you for your time reading my ebook, if you can please leave an honest review.

Alan Fredrich

www.ingramcontent.com/pod-product-compliance
Lightning Source LLC
Chambersburg PA
CBHW040327220526
45473CB00009B/2596